Cover: Alternative fuels plant in Montana
      Tactical wheeled vehicle
      An architecture for series hybrid enectric vehicle design

This page intentionally left blank

Future Fuels

April 2006

OFFICE OF THE ASSISTANT SECRETARY OF THE NAVY
(RESEARCH, DEVELOPMENT AND ACQUISITION)

NRAC 06-1

This page intentionally left blank

Table of Contents

Executive Summary

During the advance on Baghdad senior Marine and Army field commanders had many significant interdependent variables to contemplate, in addition to the basic capability and intent of the Iraqi forces before them. In order to maintain both the velocity and operational tempo of their highly mobile forces located across a wide battle space the subject of fuel was an ever present consideration. Much time, energy and continuous analysis was put into determining when, or if, a culminating point would be reached due to this vital resource. The challenge "Unleash us from the tether of fuel," came from Lieutenant General James Mattis, Commanding General (CG) of the Marine Corps Combat Development Command (MCCDC), and his Operation Iraqi Freedom (OIF) experience as CG of First Marine Division. Mattis' challenge was taken on by John Young, then the Assistant Secretary of the Navy (ASN) (Research, Development and Acquisition [RD&A]) who directed that the Naval Research Advisory Committee (NRAC) identify, review, and assess technologies for reducing fuel consumption and for producing militarily useful alternative fuels, with a focus on tactical ground mobility. Technical maturity, current forecasts of "market" introduction, possible operational impact and Science & Technology (S&T) investment strategy were considered. The principal findings of this study fall in two main time-frames.

As a near-term response to Gen Mattis' challenge, the Panel determined that the fuel tether remains, but found a way to lengthen it (Hybrid Electric Vehicle technology) and untangle it (dynamic fuel management). During (PR07/POM-08), the Marine Corps must commit to the development of the hybrid electric architecture for tactical wheeled vehicles and the development of sensor and communications systems to enable operational commanders to manage fuel allocation and re-supply in real-time during combat operations. These two near-term responses are described as:

1. Hybrid electric drive vehicles offer the most effective and efficient way to meet LtGen Mattis challenge. Improved fuel economy, as much as 20% or more, can significantly reduce the existing Marine Expeditionary Force (MEF) shortfall in fuel as well as reduce the expeditionary footprint. Hybrid electric drive vehicles enable highly maneuverable and agile vehicle traction control both on and off-road, in covert or overt operations, and can provide mobile electric power. This vehicle architecture also offers additional trade-offs in reach and mobility as related to a systems capability. To achieve improved reach and mobility, a hybrid electric strategy must be developed leveraging commercial sector and Army investments.

2. Presently the Marine Corps and the Army do not have the ability to effectively and efficiently manage fuel during combat operations. As operational reach is extended, accurate planning tools, real time vehicle level fuel status, and location data indicators are critical to enabling dynamic retasking of fuel assets on the battlefield, and to providing the ability to conserve fuel, sustain op tempo and reduce fuel train vulnerability.

In the farther time-frame, numerous alternative fuels are being evaluated across the spectrum of power and energy density to satisfy tomorrow's fuel needs for the U.S.; only liquid hydrocarbons can provide the Department of Defense (DOD) with the properties needed for its transportation fuels in the foreseeable future. Currently, these fuels are obtained from refining petroleum, but these resources are dwindling and must be replaced with a suitable substitute. Fortunately, the U.S. has large deposits of coal and shale oil, and Canada has large tar sand deposits. DOD should play an active role in catalyzing the development of this US infrastructure and ensure that it will be able to make use of manufactured fuels for its vehicles. The Panel finds that DOD needs to commit now to procuring manufactured liquid hydrocarbons for the long term at lower than current market price to encourage commercial financing, push technology and help motivate the building of the necessary manufacturing and distribution infrastructure.

Terms of Reference

During the advance on Baghdad senior Marine and Army field commanders had many significant interdependent variables to contemplate, in addition to the basic capability and intent of the Iraqi forces before them. In order to maintain both the velocity and operational tempo of their highly mobile forces located across a wide battle space the subject of fuel was an ever present consideration. Much time, energy and continuous analysis was put into determining when, or if, a culminating point would be reached due to this vital resource. From this OIF issue the Marine Corps developed the basis of the Terms of Reference (TOR) for the Future Fuels study. The critical issue of the TOR came from our sponsor, LtGen Mattis..."unleash us from the tether of fuel". This challenge also resulted in the Panel examining how to "untangle" the tether of fuel in terms of fuel utilization and management.

The challenge from our sponsor provided a lens through which the Panel was able to more clearly focus on tactical ground mobility and increased operational reach. The Panel identified, reviewed and assessed technologies that would reduce fuel demand while supporting mission objectives, consequently increasing the operational options. In this area the Panel evaluated near-to-mid term opportunities for assessment.

A final element of the TOR, with a view towards the longer term, was the examination of militarily useful alternative fuels. This is relevant on future battlefields and to our national strategy. As developing economies in Asia rapidly increase their consumption of petroleum-derived hydrocarbon fuels, they will be competing with the US, which now dominates world oil consumption. Such competition will drive prices ever higher, and perhaps lead to intermittent fuel shortages as production fluctuates. Clearly, this competition for resources also provides oil producers multiple options for selling their products, and raises the possibility that the US could face shortages resulting from shifts in political alignments within the producing nations. Furthermore, US dependence on foreign petroleum resources could result in strategic or tactical liabilities in the future.

A complete copy of the TOR can be found at Appendix A.

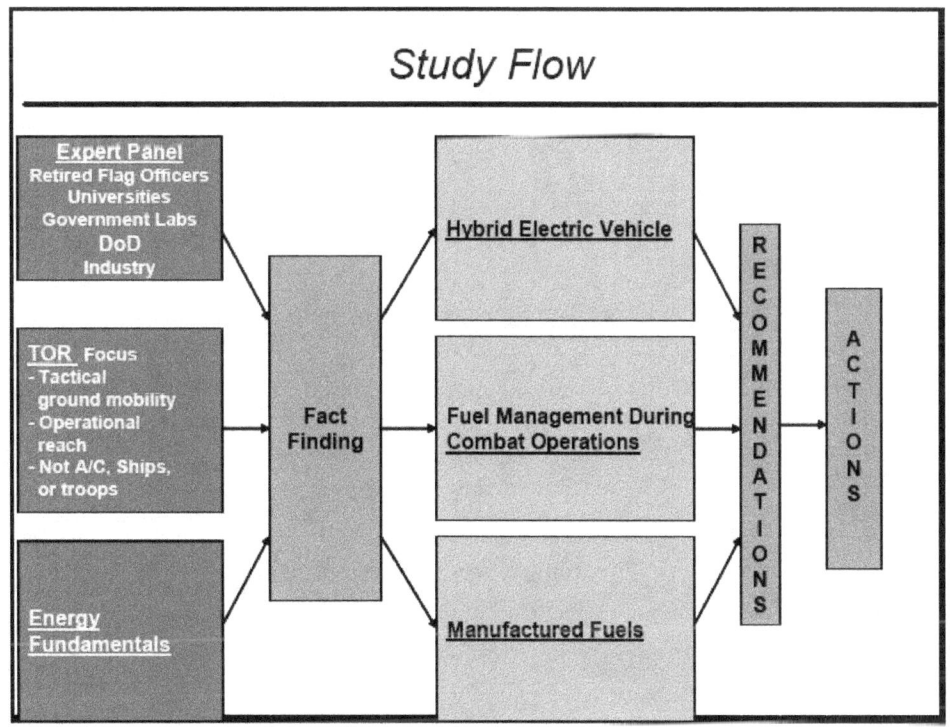

Study Flow

The Panel had a complex challenge and developed a logical approach to address the TOR. The Study Flow plan was assembled to represent this. First subject matter experts were selected from the ranks of retired flag officers, Universities, Government Laboratories, DOD, and Industry. The TOR was developed with and approved by the study sponsor, LtGen James Mattis and his staff. The focus of the Panel's work was on improving tactical ground mobility and increasing operational reach. The Panel did not study the areas of fuels for aircraft, ships or troops since each of these had been investigated and considered by others in the past.

To aid the Panel's examination of current and alternative fuel approaches and to establish metrics for usefulness, it was necessary to first understand certain energy fundamentals of transportation fuels. With these fundamentals in hand, along with the TOR and an expert panel, fact finding briefings were requested from the military services, DoE and Office of the Secretary of Defense (OSD), Universities, Defense Advanced Research Projects Agency (DARPA), the petroleum industry and the commercial vehicle industry.

These fact findings resulted in conclusions in three major areas: Hybrid Electric Vehicles (HEVs), fuel management during combat operations and manufactured fuels to address the increasing world wide demand, the decreasing supply of petroleum. From these conclusions followed Recommendations and Actions.

This page intentionally left blank

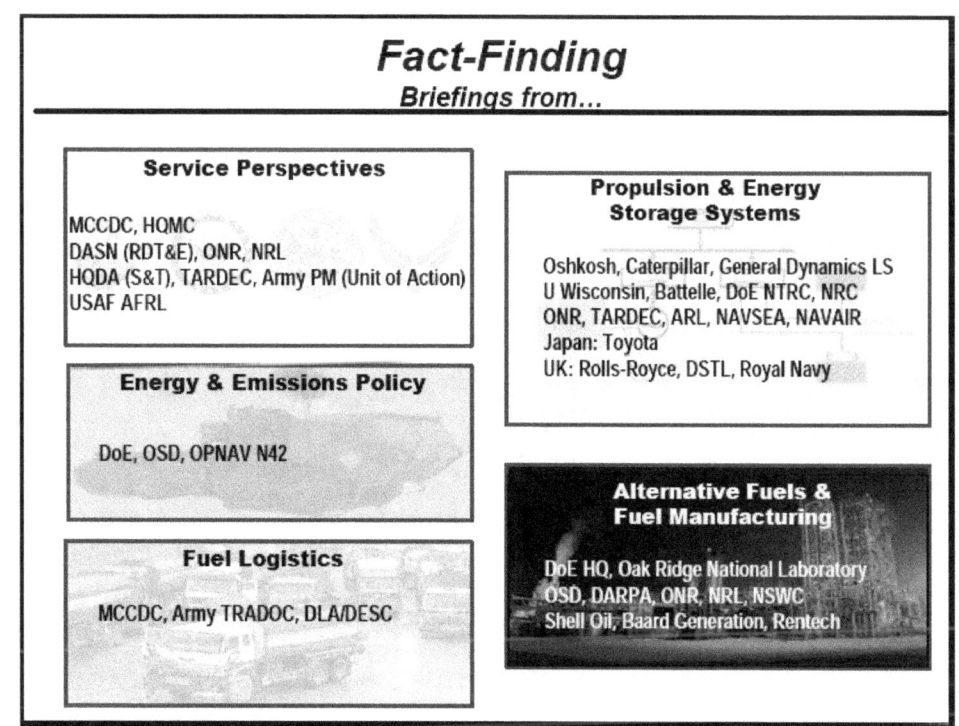

Fact-Finding

Once the energy fundamentals were understood briefings were received from the MCCDC and Marine Corps Headquarters (HQMC) to establish Marine Corps requirements, concerns and limitations. Briefings from the Deputy Assistant Secretary of the Navy for Research, Development, Technology and Evaluation (DASN (RDT&E)), the Office of Naval Research (ONR) and the Naval Research Laboratory (NRL) provided insight on Department of Navy (DoN) efforts related to the study topics. A perspective on the activities of the other services on future fuels was given by briefings from Department of the Army Headquarters (S&T), the Army Tank and Automotive Research, Development and Engineering Center (which has primary responsibility for ground vehicle technologies), Army Program Manager for Unit of Action and the Air Force Research Laboratory.

Both the Army and Marine Corps provided useful information on military fuel logistics requirements, procedures and problem areas. Briefings from the DoE, OSD, Office of the Chief of Naval Operations (OPNAV) and N42 helped the panel understand the impact of current and future emission standards as well as the likely DoN response.

Manufacturers of tactical wheeled vehicles provided insights on design of HEVs, areas for increased emphasis and strategies for the future. Toyota was extremely helpful in explaining the motivations for and likely evolution of hybrid electric passenger vehicles and fuel cell powered vehicles. Briefings from government laboratories, universities and a trip to the United Kingdom (Ministry of Defense and U.K. industry) identified research efforts in support of hybrid electric tactical wheeled vehicles.

Finally, the study panel sought information on alternative fuels and fuel manufacturing from the DoE, various DOE and DOD laboratories and representatives from the oil industry. Both Shell Oil and Bard provided insights on alternative sources of fuel.

This page intentionally left blank

Takeaways

- *Fuel Economy is Combat Power ...*
  a key performance parameter
- *Liquid hydrocarbons ...*
  the ideal transportation fuel
- *No single "silver bullet" to 50% reduction in fuel consumption*
- *Key actions:*
  - *Commit to hybrid electric architecture for Tactical Wheeled Vehicles (TWV)*
  - *Long term commitment to manufactured liquid hydrocarbon fuels from domestically abundant feedstocks*

Takeaways

As stated earlier, during the 2003 advance on Baghdad the Marine Corps and the Army had to maintain both the velocity and operational tempo of their highly mobile forces. With these forces located across a wide battle space, the subject of fuel was an ever present consideration. Much time, energy and continuous analysis was put into determining when, or if, a culminating point would be reached due to this vital resource.

Thus to ensure that operational commanders are better able to achieve their missions, system engineers and designers need to work with military users to better design future vehicles with increased fuel efficiency to maximize combat power. In order to mitigate transportation and on board storage requirements, high energy density fuels are essential. Liquid hydrocarbon fuels, such as diesel, represent the highest energy density fuels available for ground transportation. Asked to develop the ideal transportation fuel, a chemist stated that the result would be a liquid hydrocarbon.

While the panel identified no single action that would achieve the goal of reducing fuel consumption by 50%, it is clear that improving the management of fuel resources on the battlefield can lead to a significant extension of operational reach. In addition, two areas were identified for future work:

1. HEV architecture for tactical land vehicles offers improved operating efficiency while also improving mission flexibility and easing field maintenance requirements. Series HEV architecture enables all vehicles to provide electric power up to the full capacity of their engines, thus eliminating the need for separate generators and reducing overall footprint.

2. Manufactured liquid hydrocarbon fuels offer the needed energy density with attractive independence from foreign sources of petroleum. DOD could provide the catalyst that initiates a commercial market infrastructure.

This page intentionally left blank

ENERGY FUNDAMENTALS

A first element of the Panel's study was to focus on the Energy Fundamentals, identifying the critical parameters which determine the efficacy of any proposed solution. These include fuel energy density, tactical mobility design constraints, MEF fuel usage at today's optempo in OIF and in the future. These are described in the following pages.

This page intentionally left blank

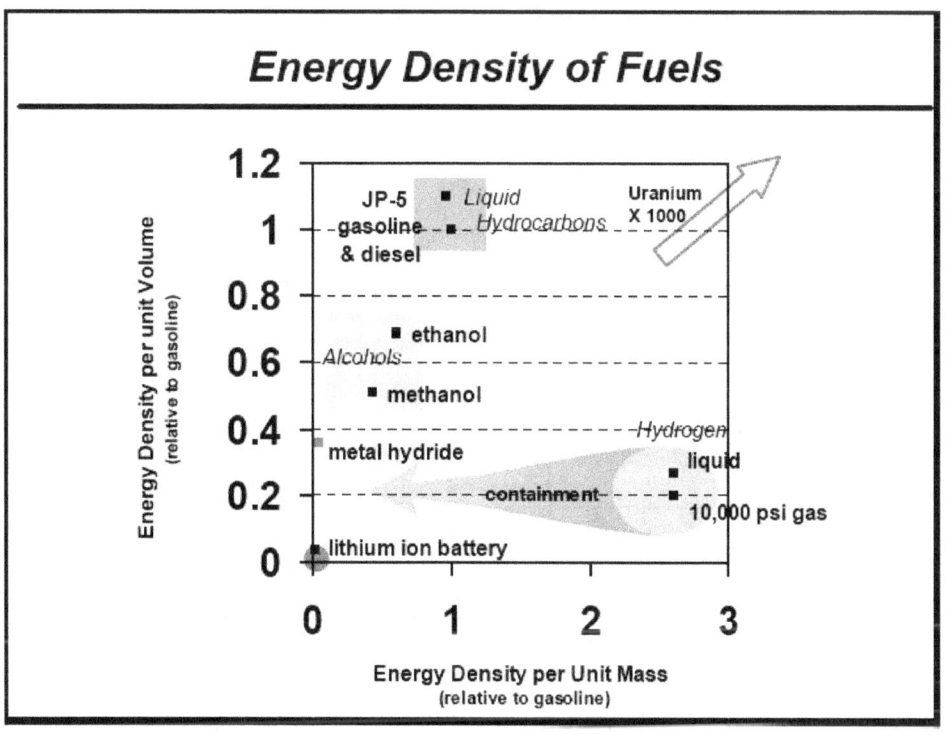

Energy Density of Fuels

Fuels may either be derived directly from natural resources (e.g. petroleum, natural gas or uranium) or by a method of storing energy in a more convenient form (e.g. alcohol from biomass or hydrogen from electrolysis of water). As such, the stored energy density is a useful metric for comparing various fuels. Since fuels may be solid, liquid or gaseous, both gravimetric (energy per unit mass) and volumetric (energy per unit volume) energy densities are important. The above chart compares the volumetric and gravimetric energy densities of liquid hydrocarbon, alcohol and hydrogen fuels along with those of batteries.

Other than uranium, the liquid hydrocarbons offer the most attractive combination of volumetric and gravimetric energy densities. The alcohols offer approximately half of the energy density of the liquid hydrocarbons. Although all of the fuels require containment, the only fuels on the chart that sustain a significant impact on energy density due to containment are the hydrogen fuels (due to the gaseous nature of hydrogen). Liquid hydrogen requires cryogenic storage at -253°C which consumes energy equal to about 30% of the energy being stored. Pressure vessels required to contain gaseous hydrogen impose a penalty of 10 to 20 times the weight of the hydrogen being stored. The impact is to move the effective gravimetric energy density of hydrogen fuels substantially to the left on the chart.

Another containment technology for hydrogen is to combine hydrogen with metals to form metal hydrides. However, the weight of the metals required and the low fraction of hydrogen stored combine to produce low resulting energy densities. Additionally, heat is typically required to release the hydrogen from the hydride when it is required. For reference, the best batteries offer energy densities 30 to 50 times lower than liquid hydrocarbon fuels.

Fuel	Energy per Unit Mass	Energy per Unit Volume
Gasoline	1.0	1.0
JP-5	0.97	1.1
Methanol	0.44	0.51
Ethanol	0.61	0.69
Liquid Hydrogen	2.6	0.27
Metal Hydride	0.046	0.36
Methane (@ 3,000 psi)	1.1	0.29
Hydrogen gas (@ 3,000 psi)	2.6	0.06
Liquid propane (@ 125 psi)	1.0	0.86
Methane (@ 10,000 psi)	1.1	0.97
Hydrogen gas (@ 10,000 psi)	2.6	0.2
Lithium ion battery	0.019	0.035

For reference purposes, the table above lists volumetric and gravimetric energies for various fuels relative to the energy density of gasoline.[1] The lithium ion battery, representing the most energy dense fielded battery technology, is included for comparison purposes.

---

[1] "Hydrogen as a Fuel for DOD", T. Coffey, et al, <u>Defense Horizons</u>, Nov. 2003

Tactical Mobility Fuel

- **Tactical Vehicle designs impose severe limitations on volume and weight**

- **Energy Density is therefore the primary figure of merit for transportation fuels**

- **Hydrogen presently unsuitable for a tactical mobility fuel**
  - *made using other fuels*
  - *containment reduces energy density a factor of 10 to 20*

Liquid Hydrocarbons are the ideal fuel for tactical mobility

Tactical Mobility Fuel

Fixed site applications, such as commercial electric power generation, are typically less sensitive to energy density of fuels than are applications like transportation. Since military aircraft and tactical ground vehicles are typically both weight and volume limited, military transportation fuels are among the most demanding applications in terms of fuel energy density. Any requirement to armor tactical ground vehicles exacerbates the limitations on fuel weight and volume. The lesson derived from this chart is that, of the available options, the liquid hydrocarbon fuels are the ideal choice for tactical mobility by a substantial margin.

Although hydrogen has exceptionally high gravimetric energy density, the containment penalty makes it unsuitable as a tactical mobility fuel. Furthermore, it is typically manufactured by thermal reforming of natural gas (methane) or by electrolysis of water. This means that other energy resources are required to produce hydrogen (as opposed to petroleum based fuels). Hydrogen produced by electrolysis using nuclear generated electricity may prove to be useful in the longer term as a feedstock for manufactured fuels.

This page intentionally left blank

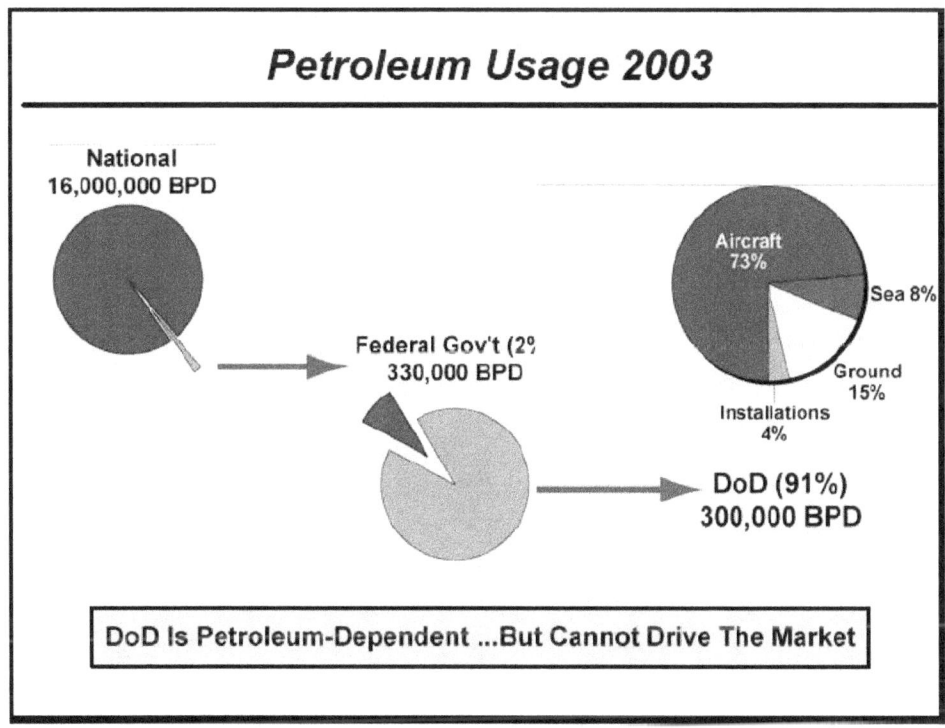

Petroleum Usage 2003

The United States consumes over 16 million barrels (almost 700 million gallons) of petroleum products per day. As a reference point China now consumes about 6.5 million barrels a day. Nationwide, about half of the US fuel is consumed in automobiles and trucks. The federal government's petroleum demand is a mere 2 percent of this total, at 330,000 barrels (near 14 million gallons) per day.

Within the federal government, the DOD is the big consumer, requiring about 300,000 barrels per day for normal peacetime operations across the four DOD Services. Aircraft consume approximately 75 % of DOD petroleum products. The Air Force continues to have primary responsibility to improve aircraft engine parameters including efficiency and the quality of fuel. Although ships use a much lower percentage of the DOD petroleum products, the US Navy continues efforts to increase power plant efficiency and fuel quality through programs like the Integrated Propulsion System (IPS). There are a number of programs within the Army and Marine Corps that are aimed at the soldier, including reducing required power and increasing energy storage and decreasing the weight of batteries.

It is clear that DOD is petroleum dependent. In weight and volume limited applications, which are typical of transportation functions, there is no substitute for the energy density provided by liquid hydrocarbons. That said, DOD cannot drive the market due to its relatively minor portion of the national petroleum demand. While its position is not one of strength, it has some opportunity to influence the market through purchase guarantees, for example. If it is to ensure an adequate supply of liquid hydrocarbon fuels for its operations into the future, DOD must look to innovative arrangements that will stimulate supply from domestic resources.

This page intentionally left blank

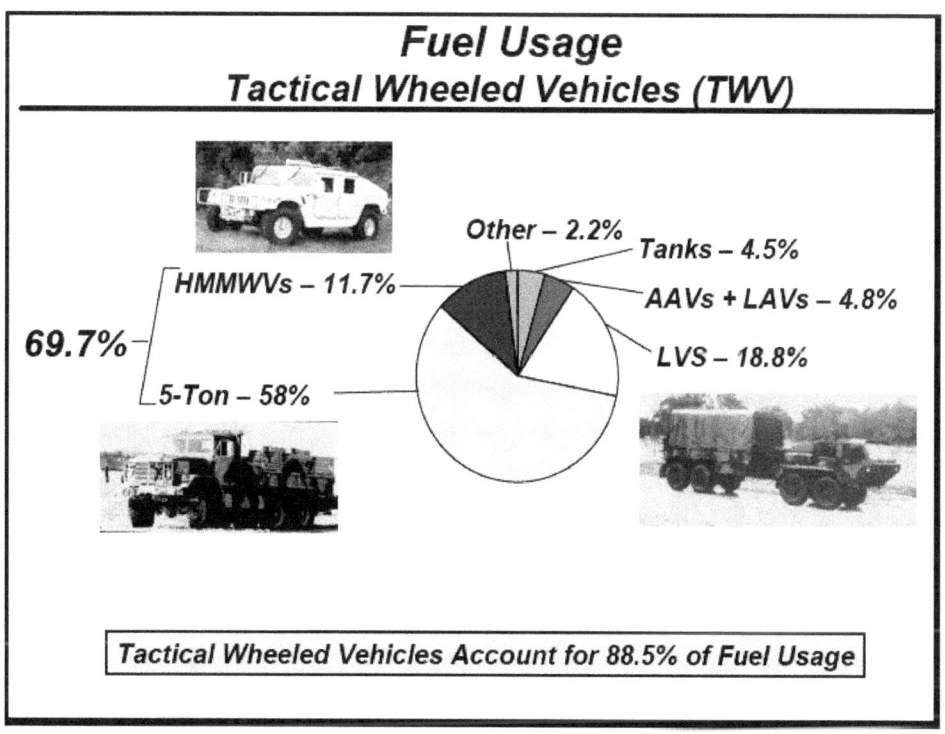

Fuel Usage

Data from the Marine Corps' 2003 MEF Fuel Use Reduction study shows that almost 90% of the fuel used by MEF ground vehicles will accrue to TWVs, including High Mobility Multi-Purpose Wheeled Vehicles (HMMWVs), Medium trucks, and the Logistics Vehicle System (LVS). Moreover, the study shows conclusively that the armored weapon vehicles, e.g., the M1 tanks, Light Armored Vehicles (LAVs) and Amphibious Assault Vehicles (AAVs), although fuel guzzlers individually, as a fleet these vehicles consume a relatively minor fraction of the fuel apportioned to MEF ground vehicles.

Consequently, TWV should be the primary target for fuel economizing. With realistic projections of fuel economy improvements on the order of 20% for series hybrid electric propulsion schemes in TWV, the fuel savings across the MEF TWV fleet can be upwards of 50,000 gallons per day, all of which can be applied to improve reach and speed of the maneuver force.

This page intentionally left blank

```
┌─────────────────────────────────────────────────────────┐
│  ┌───────────────────────────────────────────────────┐  │
│  │    TWV Operational Tempo and Mission Profile      │  │
│  │  ─────────────────────────────────────────────    │  │
│  │                                                   │  │
│  │  • May 2005 IGMC Findings from OIF:               │  │
│  │                                                   │  │
│  │    – "The fleeting nature of insurgents demands   │  │
│  │       highly responsive, highly maneuverable and  │  │
│  │       highly agile ground combat power"           │  │
│  │                                                   │  │
│  │    – All classes of TWV's average 70-75% off      │  │
│  │       road/unimproved roads                       │  │
│  │                                                   │  │
│  │    – Heavy reliance on Mobile Electric Power      │  │
│  │       (MEP) throughout the AOR                     │  │
│  │                                                   │  │
│  │  • Distributed Op's further complicate TWV        │  │
│  │    power & fuel                                   │  │
│  │                                                   │  │
│  │  • Electrical power requirements growing rapidly  │  │
│  │                                                   │  │
│  │  ┌─────────────────────────────────────────────┐ │  │
│  │  │ Future TWV's...off-road fuel efficient with │ │  │
│  │  │ power generation                            │ │  │
│  │  └─────────────────────────────────────────────┘ │  │
│  └───────────────────────────────────────────────────┘  │
└─────────────────────────────────────────────────────────┘
```

TWV Operational Tempo and Mission Profile

Military operations place demands on vehicles that are uniformly unique. Unlike commercial automobiles and trucks that have been optimized for use on highways, tactical wheeled vehicles are required to travel virtually every type of terrain, from interstate freeways to desert sands and forest or agricultural land. A single mission profile for a TWV may include segments of each, enroute to its mission responsibility.

Terrain has a direct correlation to the relative fuel economy of a wheeled vehicle. To optimize fuel economy across the range of topographies that might be encountered, the vehicle propulsion system should have the capability to adapt to a specific topography for optimized fuel performance. Clearly a design goal for hybrid electric drive technologies must be on-board/on-the-fly mission profile re-selection capability. The principle function would be for the optimization of fuel economy; however additional military relevant features such as silent watch and a stealth mode would be included.

TWVs of the future may very well be subject to continually more demanding emissions standards, both in the U.S. and in countries where U.S. forces deploy or train. These stricter standards tend to have a negative effect on the types of performance that are militarily relevant and include increased thermal signature and in some cases, reduced fuel economy. A peace time – war time emissions design architecture may also be necessary to comply during normal training, but achieve optimal performance during combat operations.

Lastly, the demand and emerging feasibility of unmanned ground vehicles is growing. HEV architecture will provide the same benefits to unmanned vehicles as it brings to TWV.

This page intentionally left blank

Findings

Future battlefield mobility requires effective utilization of fuel

- **Nearer-term payoff (PR 07/POM 08)**
  - **Vehicle architecture implementation**
  - **Commander's fuel management**
- **Longer-term payoff (2015 & beyond)**
  - **Fuel manufacturing**

Findings

The principal findings of this study fall in the two main timeframes: Nearer-term and longer-term. In the near-term, the two findings that can and must be acted on as the 07 budget is completed (PR07) and then within the POM-08 cycle to achieve payoffs in the relatively near term, are:

- Hybrid Electric Vehicles: The development of and commitment to hybrid electric architecture for TWVs. Important features are vehicle architecture, leveraging other technology investments, energy conversion options, fuel quality issues, emission standards and energy storage.

- Fuel Management during Combat: The development of sensor and communications systems, along with resource allocation tools to enable operational commanders to manage fuel allocation and re-supply in real time during combat operations. Timely delivery of fuel is essential to maintaining operational tempo. Fuel management during combat operations can include: location and fuel status of vehicles, ability to dynamically relocate fuel assets to areas of high need, etc.

This page intentionally left blank

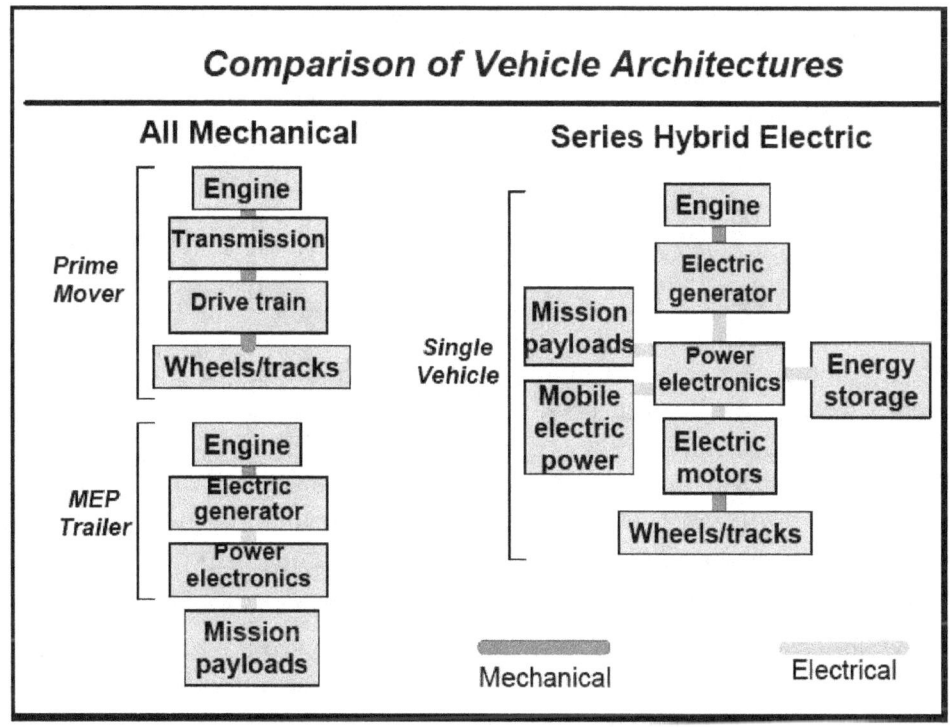

Comparison of Vehicle Architectures

In order to illustrate the differences in complexity, weight, duplication of functions and performance, the Panel compared existing all-mechanical vehicle architectures and series hybrid vehicle architectures. Comparisons are illustrated in the figure based upon reference to a tactical field unit consisting of self-propelled transport capability and with capability to provide auxiliary electrical power for a mission payload element (e.g.,a radar sensor unit).

All-Mechanical: For the current all-mechanical vehicle architecture this tactical field unit would typically consist of a medium truck pulling a trailer mounted mobile electric power unit. As can be seen from the functional diagram of the all mechanical vehicle architecture the engine power source function is duplicated for both the transport vehicle and the Mobile Electric Power (MEP) unit. The truck itself also has a very heavy and inflexible mechanical clutch, transmission and drive train assembly for transferring power to the wheels (tracks). This fixed mechanical drive train assembly is also not amenable to providing variable height ground clearance as a function of terrain.

Mechanical drive trains tend to be very reliable but when problems or battle damage does occur, they are extremely cumbersome to repair or replace. Also, as was stated earlier, the weight of these components is a major factor in limiting the overall fuel efficiency of the vehicle. In the all-mechanical architecture the mobile electric power unit (trailer mounted) has its own separate engine and generator for providing auxiliary power. It also has its own separate wheel and axle assembly as well as a trailer frame, all of which adds to the overall weight of the unit.

Series Hybrid Electric: For hybrid electric vehicle architectures, there are two main configurations: 'parallel' and 'series'. The 'parallel hybrid' (not shown) has a conventional engine/transmission connected mechanically to the driven wheels or tracks (as in the All-Mechanical above); but into this driveline an electric generator is inserted such that it also drives through the mechanical transmission system. This offers an alternative drive path and the vehicle can be driven either mechanically alone, or electrically alone or using both power inputs. In contrast to the all-mechanical approach and the parallel hybrid, the series hybrid vehicle architecture utilizes a single engine power source and a single electric generator which provides all power for both vehicle transport (propulsion) as well as for auxiliary MEP. This hybrid architecture no longer requires the use of very a heavy mechanical clutch, transmission and drive train and allows the engine to operate at ideal speed and duty cycle independent of vehicle speed thereby significantly improving fuel efficiency. This approach does allow the use of in-hub electric motors to provide mechanical power to the wheels (or tracks).

Such motors are well proven and relatively reliable but of course have much less of a track record than do mechanical drive trains. The use of in-hub electric propulsion motors requires more sophisticated power control and conditioning electronics to provide for differential power distribution amongst wheels, but offers much more flexibility in terms of closed-loop control of power at the wheels (or tracks) for improved traction and handling. This approach also offers a great deal of flexibility in terms of providing for variable height ground clearance in real-time as a function of terrain.

The series electric vehicle architecture does require a new function not required by the all mechanical architecture and that is the electrical energy storage function. This would typically take the form of a high energy density battery which would incur some additional weight for the vehicle.

All-in-all, the series hybrid electric vehicle would offer the advantages of reduced overall vehicle weight, less functional redundancy, more flexibility in performance, better fuel efficiency, and better potential for functional modularity and simplicity of maintenance.

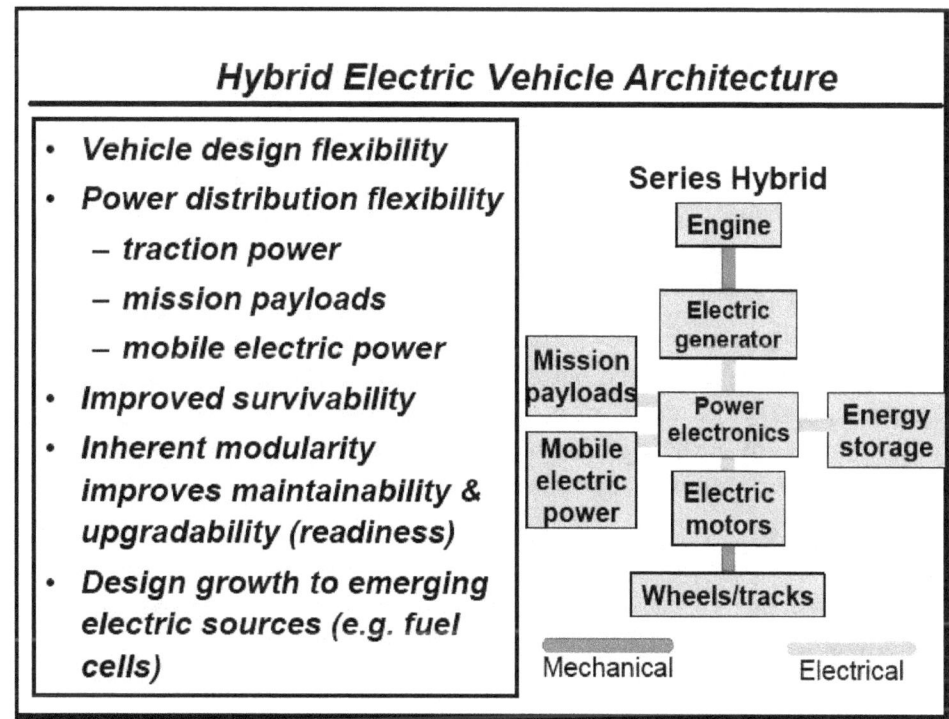

Hybrid Electric Vehicle Architecture

To obtain the maximum life-cycle cost and performance benefit from hybrid electric technology, the Marine Corps should strive to achieve a common, modular, "series" hybrid electric system architecture for all of its tactical wheeled vehicles. This architecture would utilize a standard electrical "bus" which would provide a common power transfer infrastructure for all power sources and power user functions on the vehicle. This power bus would operate at standardized voltage and would provide a power backbone into which power sources and power consumer components could be interfaced in a "plug-and-play" fashion as required for different mission configurations.

Component elements of this architecture would include primary power sources which would initially be diesel-electric generator sets, distributed electric motors at the drive wheels for propulsion and braking, as well as on-board weapon systems, sensor systems, and communications systems modules. Such a standardized common power structure would also provide an extensible framework into which new technologies could be integrated as they became mature. For example, hydrogen fuel cell power sources could eventually replace or supplement diesel electric generator sets as the primary vehicle power source and could be easily interfaced into the same electric power distribution backbone. The modular, "plug-and-play" operating characteristics of the HEV architecture also lead to improved maintainability and survivability.

A series hybrid electric architecture of the type described above would provide the greatest flexibility for vehicle design since much of the space and weight consuming aspects of conventional mechanical power distribution systems, i.e. drive shafts and transmission/differential gear boxes can be eliminated. This provides much more flexibility in terms of integration of required payload and mission packages. In addition, the series HEV architecture provides "exportable" mobile electric power as an integral part of the vehicle using the same common electric power infrastructure.

29

TWVs are not able to utilize stored energy devices to reduce propulsion engine size to the same extent as civilian vehicles. A military vehicle must be capable of continuously generating the maximum power required for mobility, unlike a civilian vehicle (e.g. hybrid drive car), as it may be necessary to move through terrain such as deep sand, which demands maximum power, for long periods and a reduction in performance when the stored energy is depleted would be unacceptable. However other systems under consideration today also require stored energy devices – intermittent loads such as electromagnetic guns, directed energy weapons and silent watch capability.

Additionally, stored energy would be useful to boost acceleration for short periods. It is possible to extract (for a limited period) without damage, additional power from an electric motor by providing additional current. It is also possible to drive the vehicle (for a limited distance and speed) by utilizing the on-board stored energy without using the engine. This provides a stealthy movement capability.

During normal driving the transient fluctuation in power levels is accommodated by drawing energy from or re-charging the energy storage system. Normal braking can be augmented electrically by using the electric drive motors as generators with the energy produced either used to re-charge the battery, or if that is fully charged, dumped into a resistor bank.

Finally the vehicle on-board power generation and energy storage system can be utilized for powering off-board systems, such as command & control systems, engineer power tools, etc. This could eliminate or reduce the number of towed generators required on the battlefield thus further reducing the logistic demand.

Opportunities to Leverage Technology
Hybrid Electric Vehicles

Technology/Action	Commercial	Army	Needed (Naval)	
			Fund	Adapt
Systems Engineering	?	?	?	
Power Electronics and Controls				
– Size		?		?
– Thermal Management		?		?
Energy Storage				
– Batteries	?	?		?
– Ultra-Capacitors	?	?		?
– Flywheels		?		?
Energy Conversion				
– Engines	?			?
– Fuel Cells	?			?
– Reformers and Desulfurization		?		?
Motors				
– Permanent Magnet		?		?
– Wound Rotor	?		?	
Series Architectures and Integration				
– Modeling and Simulation		?		?
Active Heavy-duty Suspensions		?	?	
Integration of Mission Systems				
– Weapons and Armors		?	?	
– Pulse Power Technology			?	?
Mobile Electric Power		?	?	
RST-V Demonstration			?	

Leveraging Technology

The transition of future TWV to hybrid electric propulsion architectures will require some amount of development of the component technologies. While hybrid electric propulsion in vehicles has been given a substantial boost by its application in automobiles, this market is directed largely at parallel architectures, which maximize the advantages for on-road applications. TWV, on the other hand, need both on-road mobility and off-road mobility and power generation capabilities.

The Army is addressing several of the open issues that will affect military series architecture vehicles, but there are still gaps to be filled. Some require investment; many more must be monitored at least to ensure that the developmental path being followed will yield components and technologies that fulfill the mission requirements of the Marine Corps.

The ultimate application of hybrid electric technology to Marine Corps TWV will result from the smart adaptation of commercial and Army technologies as well as the development of some of the component and integration technologies and architectures within the Naval S&T community. Organic Naval S&T support is required in the following areas:

- Systems Engineering – The need for systems engineering in all manner of development efforts has been a recurring theme in many recent NRAC studies. Too often, the S&T effort focuses on the components to the exclusion of the ways in which they will be integrated into the final design to meet the mission requirements. Establishing and following a comprehensive systems approach, from requirements through final design, must be taken as an imperative that cannot be outsourced. Early and substantial involvement of Naval S&T in the systems engineering process is essential.

31

- Wound Rotor Propulsion Motors – Most of the S&T being done today on propulsion motors in the size range relevant to TWV is focused on permanent magnet motors, often wheel mounted. Wound rotor motors provide substantially different design and control options than permanent magnet motors, at the expense of additional cooling requirements. Their benefits include higher peak torque and less sensitivity to overload. They may be more appropriate for some high performance TWV applications.

- Active Heavy-Duty Suspensions – Active suspensions have been shown to increase vehicle mobility (speed across terrain) significantly while reducing peak vertical accelerations on crew and payload. This leads to reduced fatigue, extended ability to operate, and improved durability of equipment. While the Army is supporting work in this area, the criticality of the suspension system to the fighting ability of the force makes it a candidate for closer organic attention.

- Integration of Weapons and Armor Systems – The Marine Corps has traditionally taken an active role in the development of its principal combat systems in order to ensure that the systems are effective in the unique Marine combat environment. Weapons and armor are becoming ever more sophisticated as DOD moves from the classical propellant/projectile weapons and passive armor systems to high energy weapons and active armor systems. Complexity invariably increases as the limits of effectiveness are challenged. Consequently, this is an area that requires active involvement of the Naval S&T community.

- MEP – The fastest growing requirement on the battlefield is electric power. From the power requirements of the individual Marine to the increasing power requirements for sensors, weapons, and armor systems, the need for ubiquitous electric power as the force maneuvers to its objective is burgeoning. The current solution is towed generators, which literally doubles the number of wheeled equipments that must be accommodated by the logistics system as well as the tactical vehicle fleet. Effectively making the towing vehicle the generator, due to its ability to shift its propulsion electric power to conditioned field-usable electric power, cuts the number of systems on the battlefield, simplifying the logistics and operational problems. This is an area of great potential that deserves direct involvement from Naval S&T.

- Reconnaissance Surveillance Targeting Vehicle (RST-V) Demonstration – The RST-V, a DARPA/Marine Corps project, incorporates many of the attractive characteristics of hybrid electric TWV noted above. The Naval S&T community should take the initiative to demonstrate its advantages soon.

Energy Conversion
Diesel Engines

- **Most fuel efficient**

- **Commercial engines (or derivatives) offer the most affordable choice**

- **But... commercial sector emphasis on emissions reduction leads to problems by 2010**
  - *Performance and RAM-D sensitivity to substandard fuels*
  - *After-treatment emission control systems cause significant vehicle integration and signature issues*
  - *Increased importance of emissions waiver*

Energy Conversion

One of the critical components of future series hybrid electric propulsion systems is, obviously, the energy conversion equipment. While many alternatives, including diesel engines, gas turbines, and fuel cells, have been proposed for the application, diesel engines will continue to be the best alternative for TWV propulsion for the foreseeable future. This is due primarily to the significant advances in diesel engine technology over the past several years as well as the very attractive cost of diesels, both on procurement cost and total ownership cost bases.

However, the emissions reduction requirements for diesel engines in the near term will impact engines procured for military service in several ways. The after treatment systems that will be required after 2010 require very low sulfur fuel (<15 ppm sulfur). Yet, the data for 1997 fuel procurements DOD-wide show that very little met this standard. This is especially problematic in many of the current theaters of interest, where sulfur contents as high as 3000 ppm are encountered. The post 2010 diesel after-treatment systems are also likely to be large (approximately equal to engine in size) and require an approximately 30% larger cooling system. They are also likely to have significant thermal signature. This can have significant negative impact on TWV design and performance.

While using emission-friendly engines is attractive from a cost and conformance to Environmental Protection Agency (EPA) standards perspectives, this could present problems when the vehicles are placed in theater for their primary function. A waiver from the 2010 emissions standards for TWV is critical to the Marine Corps. This issue requires the immediate attention of the hierarchy within the Marine Corps, DoN, and OSD.

33

This page intentionally left blank

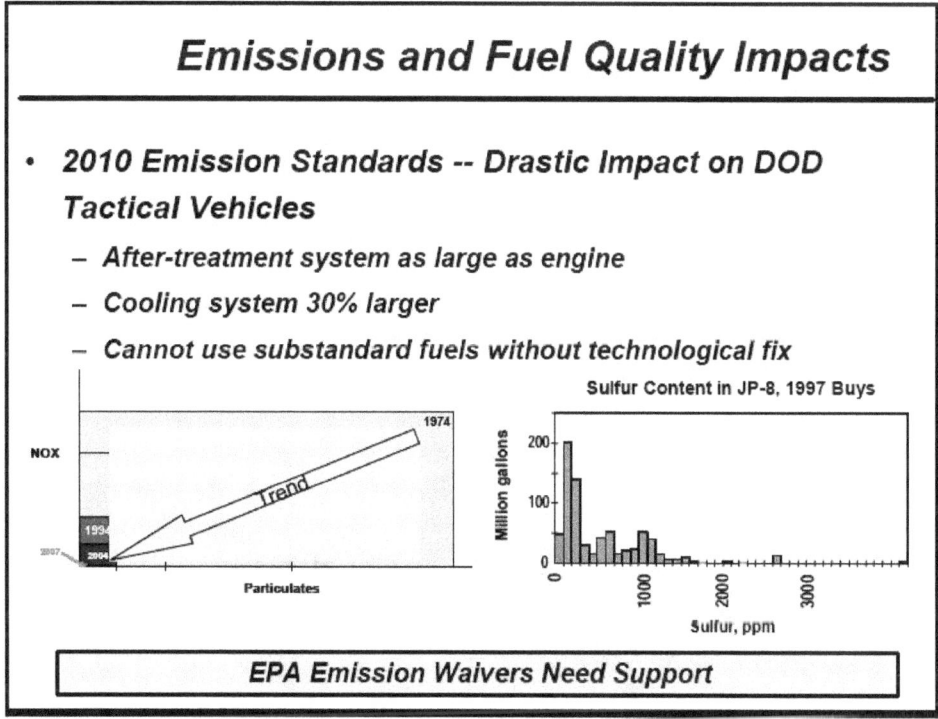

Emissions and Fuel Quality Impacts

All DOD ground combat vehicles are automatically exempt from emissions controls under 40 Code of Federal Regulations (CFR), 89.908. Examples of such ground combat vehicles are M1 Abrams tank, Bradley Fighting Vehicle, and Stryker Vehicle.

Maximum allowable emissions in 2007 and 2010 are extremely stringent for TWVs. The 2007 emissions limit is the "red box" (in the graph), while the 2010 emissions box would essentially be a "dot" at the graph's ordinate. If military ground tactical vehicles are required to meet 2007 and beyond EPA emissions regulations very adverse effects occur which result in: 1) a 30% increase in cooling system size and power consumption, 2) large engine system volume, cost, and signature increases, and 3) significant system fuel economy deterioration. These adverse effects primarily result from the complex after treatment and cooled exhaust gas recirculation (EGR) systems required to meet these emissions requirements.

DOD already has some emissions waivers approved on a case-by-case basis for military tactical vehicles. Also, DOD is near official approval for a blanket EPA emissions waiver for tactical military vehicles concerning future emissions regulations (including 2007 and beyond). It should be noted that DOD has always made a best case effort to meet emissions standards (over 90% reduction since 1980). However future standards would have drastic negative effects on DOD tactical vehicles without EPA waivers on a blanket basis for emissions standards for year 2007 and beyond.

This page intentionally left blank

Fuel Cells
Long Term Alternative to Engines?

- **Potential benefits**
  - *Efficiency*
  - *Pollution free, low signature*
  - *Electric power availability*
- **Commercial sector**
  - *primary source of technology for vehicle applications*
  - *focused on hydrogen fuel*
- **Military use: diesel fuel reformer / desulfurizer development critical**
- **Technical challenges include:**
  - *power density*
  - *cost*
  - *low temperature operation*
  - *start-up time, throttle response*
  - *durability*

> **Not required for hybrid electric vehicles**

Fuel Cells

Fuel cells may offer an attractive alternative to diesel engines in the long term. Their benefits include good operating efficiency, low pollution and low acoustic and thermal signature. They produce electricity directly (no generators needed) and discharge pure water as well. Since fuel cells are key to transportation in a hydrogen economy, DOE and industry are vigorously pursuing fuel cells for passenger vehicle usage. Industry is also examining fuel cell application for auxiliary (idle) power units in long-haul truck cabs.

Fuel cells require high purity hydrogen. Using logistics (hydrocarbon) fuels means that contaminants, including sulfur which poisons fuel cell catalysts, must be removed before the "clean" hydrocarbon is broken down to produce hydrogen for fuel. Reformers required to make hydrogen are presently bulky and energy intensive. Work is required to develop desulphurizers and reformers suitable for use in TWV.

Current fuel cells for vehicles are: relatively bulky, have low power density, require precious metals that are expensive, and cannot be started up or shut down as quickly as diesel engines. Improved power density and tolerance to contaminates are crucial to TWV use. Existing fuel cells are developmental models so durability is unknown. Research to address these shortcomings is underway, and the DoN should stay abreast of developments. Although the HEV architecture for TWVs does not require fuel cells to be attractive, TWVs will directly benefit from fuel cell technology when it becomes available, providing that the diesel desulphurization and reforming issues have been addressed successfully.

This page intentionally left blank

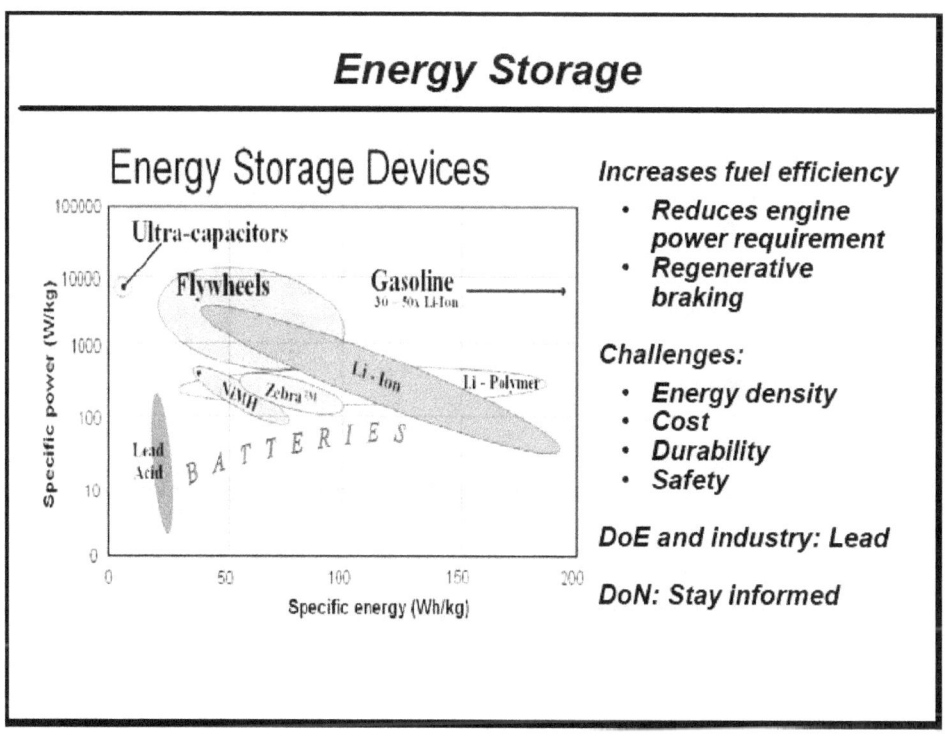

Energy Storage

A hybrid electric architecture enables prime movers to be sized to meet average rather than peak power requirements **IF** energy storage systems can supplement the engine to meet peak demand. Energy storage is also necessary for regenerative braking and silent watch functions. The chart above shows the energy and power densities of current energy storage devices. Generally, batteries offer better energy densities while ultra-capacitors and flywheels provide higher power densities. All are substantially inferior to liquid hydrocarbon fuels for bulk energy storage. Batteries are useful in HEVs for energy intensive functions like start-up, silent watch and stealthy movement while flywheels and ultra-capacitors will find application for more power intensive functions including regenerative braking, active suspension and electric weapons.

Advancements in "vehicle-sized" devices are important to future HEV transportation. There is currently substantial DOE and industry effort on this, but many challenges remain including energy density, cost, durability, and stability. (Energy dense storage meeting consumer hazard and durability standards is important to mass transportation uses, and low cost is essential.)

DoE is investing in research to meet the needs of mass transportation. Industry is developing more near-term applications. To apply scarce resources only where they are needed, the DoN should consider a formal group to follow DOE and industry's progress on HEV technologies with an eye on adapting commercial products for naval uses.

This page intentionally left blank

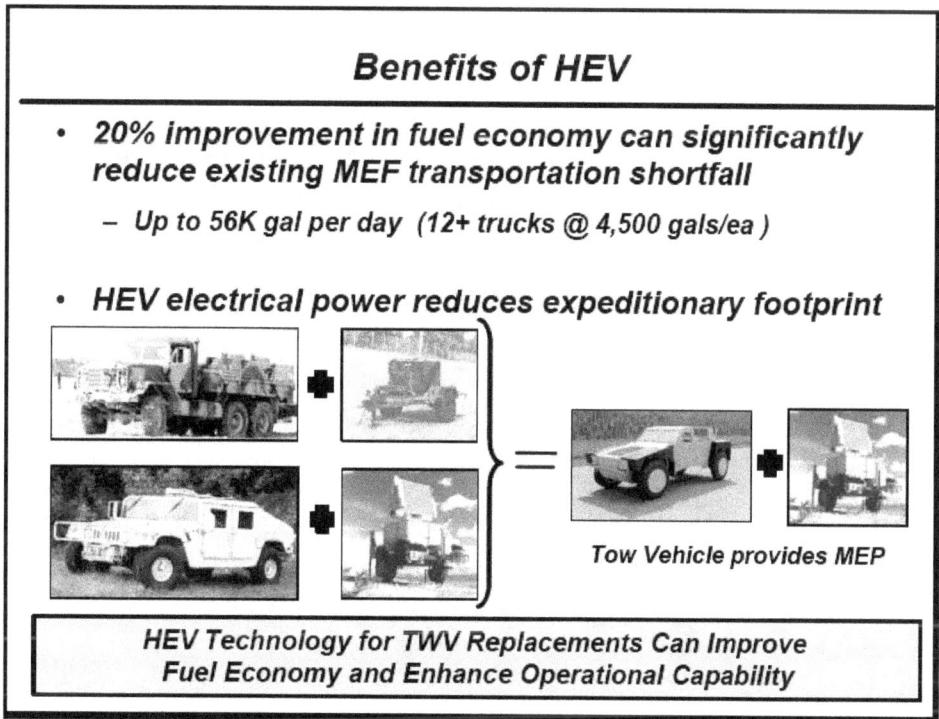

Benefits of HEV

- **20% improvement in fuel economy can significantly reduce existing MEF transportation shortfall**
  - *Up to 56K gal per day (12+ trucks @ 4,500 gals/ea)*

- **HEV electrical power reduces expeditionary footprint**

Tow Vehicle provides MEP

HEV Technology for TWV Replacements Can Improve Fuel Economy and Enhance Operational Capability

Benefits of HEV

The use of a standardized series hybrid electric architecture for all of the Marine Corps TWVs can provide a number of life-cycle operational benefits. Most importantly the nominal average, across-the-fleet, fuel savings of approximately 20% achievable through HEV technology can help reduce shortfalls in fuel supply – shortfalls which have been experienced by the MEF during combat operations. The Panel estimate that, on average, the resulting savings in fuel usage on a daily basis for the HMMWV and medium truck fleet could be up to 56,000 gallons per day. The logistics supply impact of this savings would be equivalent to eliminating the need for 11 fuel truck re-supply loads at 5,000 gallons each.

The inherent ability of the HEV to provide exportable mobile power can greatly reduce the footprint of the MEF by eliminating the need for a large number of current trailer mounted mobile electric generator sets as well as many of the medium trucks or other vehicles which are dedicated to towing these in the area of combat.

In addition, the cascading effects of a decreased fuel logistics train will yield even more tactical benefits. These include the reduced exposure of the force to attack at fueling points, the reduced need for towed generator systems due to the hybrid electric vehicles' ability to provide off-board electric power, and fewer helicopter fuel logistics sorties. There is a strong probability that secondary and tertiary benefits like this will more than make up for the direct economies of the TWV themselves, thus enabling a net effect of improved operational reach and speed.

This page intentionally left blank

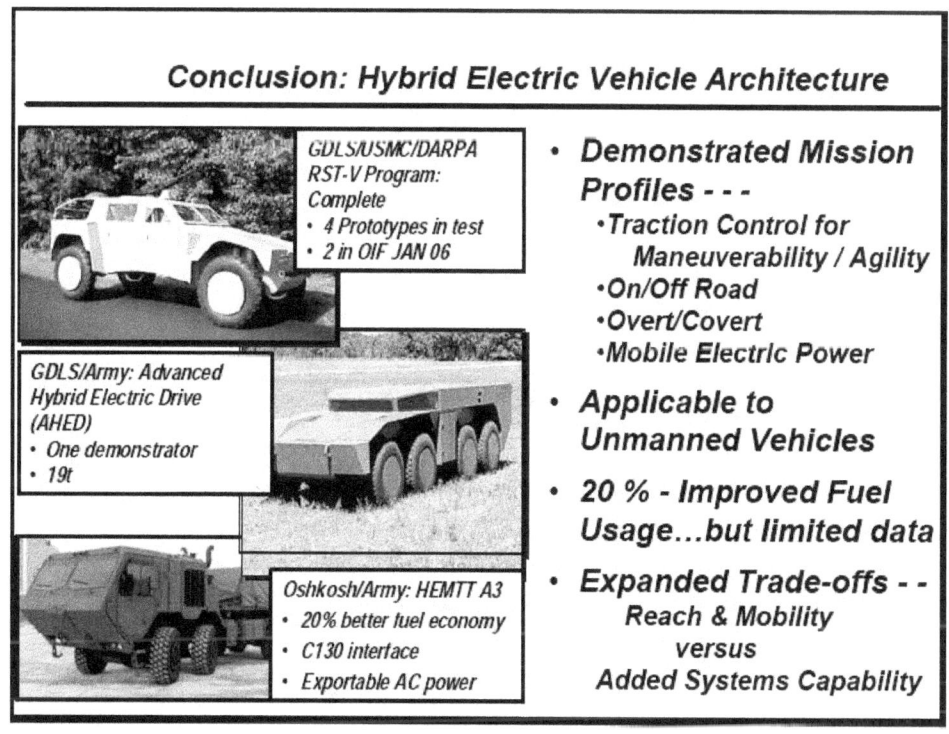

Conclusion: Hybrid Electric Vehicle Architecture

Conclusion: Hybrid Electric Vehicle Architecture

Prototype HEVs have been successfully developed and demonstrated for a variety of tactical military mission profiles. Three of these are shown in the above figure. General Dynamics Land Systems (GDLS) together with USMC and DARPA have developed and demonstrated the prototype RST-V. This is a HMMWV-type vehicle, for which plans exist to introduce two into OIF in January, 2006.

Similarly GDLS and the Army have successfully developed and demonstrated a larger 19 ton vehicle called the Advanced Hybrid Electric Drive (AHED) vehicle.

Finally, the Army and Oshkosh Truck have successfully developed and demonstrated the Heavy Expanded Mobility Tactical Truck (HEMTT) A3 heavy truck system using all hybrid electric technology, not only for propulsion but also as high-capacity exportable power source.

The potential fuel savings available through the use of HEV technology is very dependent upon the mission profiles for the vehicle operation. In reviewing much of the available HEV test results for various mission profiles, the panel found that fuel savings ranged between 5% and 55% depending upon usage conditions. Based upon this, the Panel concluded that a conservative number for average fuel savings available through the use of hybrid electric technology is on the order of 20%. Finally it is worth noting that much of the emerging hybrid electric technology could also be applied to unmanned vehicle missions with similar resultant average savings in fuel usage.

It is important to realize that any such savings in fuel consumption is then available to the tactical commander in the field for use as the situation demands. Given the same vehicle capabilities, the additional availability of fuel can be used to expand his combat reach and maneuverability. On the other hand, the added fuel may be used to support additional power requirements associated with incorporating additional weapons and sensor systems rather than in extending operational reach. These important types of trades, between operational reach and maneuverability versus additional system capability, will have to be carefully considered by the tactical commander.

```
┌─────────────────────────────────────────────────────────┐
│          Fuel Management During Combat Operations        │
├─────────────────────────────────────────────────────────┤
│                                                          │
│   •  Improved fuel management increases operational reach│
│                                                          │
│   •  Comprehensive fuel visibility/dynamic allocation:   │
│       – Conserves fuel and sustains op tempo             │
│       – Reduces the number/vulnerability of fuel trains   │
│                                                          │
│   •  Marine Corps' macro fuel estimating tool needs two  │
│      additional critical elements                        │
│       – Automated vehicle fuel status and location reporting│
│       – Dynamic tasking via Blue/Red/Terrain data fusion │
│                                                          │
│   ┌──────────────────────────────────────────────────┐  │
│   │   Fuel is not simply a commodity or logistics issue – │
│   │        it is an operational imperative            │  │
│   └──────────────────────────────────────────────────┘  │
└─────────────────────────────────────────────────────────┘
```

Fuel Management During Combat Operations

In addition to developing and fielding vehicle technologies that conserve fuel to extend operational reach, another fuel conservation measure is potentially available for use by the ground forces. The fact findings resulted in conclusions in three major areas; the second of these is the need for fuel management during combat.

The distribution of the fuel itself is an important consumer of available petroleum stocks. Timely delivery is essential to help maintain operational tempo. Fuel management during combat operations is a vital function contributing to tactical success. To deliver fuel in the most efficient and timely manner to dispersed units across the battle space several fundamental elements of information must be known. These include the location and fuel status of each tactical vehicle including all types of refueling assets, the location of both friendly and enemy forces, and a detailed knowledge of the terrain in the Area of Responsibility (AOR).

Other factors are also necessary for dramatically improved fuel management, but the most fundamental are those mentioned above. The ability to see in real time the fuel picture of all assets in the battle space, combined with the ability to dynamically reallocate petroleum assets as combat operations evolve can greatly improve the efficient delivery of this scarce and critical resource.

In addition to contributing to sustained operational tempo and extending operational reach, the number and frequency of fuel trains/sorties could also be reduced, with a corresponding reduction in the vulnerability of these assets and the number of Marines pulled from combat units to protect them. A macro fuel estimating tool is important but is not enough. The Commander also needs to possess an automated fuel status and reporting system down to the individual vehicle level, and a dynamic tasking capability that is able to fuse the friendly, enemy and topographic picture of the battle space. These capabilities when combined will further extend the operational reach of the Ground Combat Element (GCE) within the fixed quantity of fuel they currently possess.

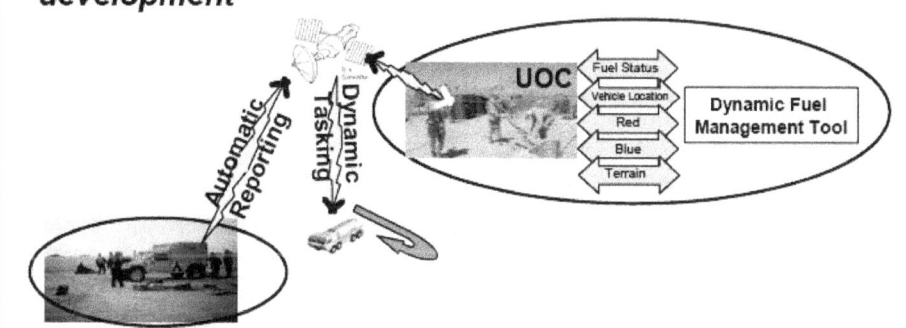

Conclusion: Fuel Management

To substantially improve fuel management during combat operations, an integrated system of new hardware and software tools need to be introduced into the GCE.

During the fact-finding phase, the study panel became aware of an ongoing project within the Marine Corps evaluating a specific technical approach on fuel management. These activities should be supported, and the field of evaluation expanded. Application to all mobility assets of the GCE must be included and not limited to only fuel transportation systems. A dynamic allocation system includes the automatic vehicle location/fuel status reporting segment but goes a considerable step farther. A complete fuel management system must include at a minimum, the ability to fuse the friendly and enemy situation, as well as integrate the topography of the AOR. These are the critical parameters necessary to properly create and evaluate real time fuel reallocation courses of action. The dynamic allocation system should have the ability to create these initial courses of action for evaluation by the Commander and his staff. The Panel recommends that the Marine Corps should not wait to pursue and field these two activities until the larger "autonomic logistics" effort is complete, but rather integrate these efforts as modules into the autonomic logistics system (when it is eventually fielded).

A near term opportunity is found in the automatic fuel status reporting requirement. Commercial fuel reporting systems like those found in the trucking and railroad industries may serve as an initial model to be adapted for military use.

This page intentionally left blank

Findings

Future battlefield mobility requires effective utilization of fuel

- **Nearer-term payoff (PR 07/POM 08)**
  - *Vehicle architecture implementation*
  - *Commander's fuel management*
- **Longer-term payoff (2015 & beyond)**
  - *Fuel manufacturing*

Findings

The principal findings of this study fall in two main time-frames. The nearer-term was covered in the previous sections, the longer-term is presented here.

In the longer-term timeframe, commercial infrastructure will allow the manufacture of high-quality transportation fuels from plentiful domestic feed stocks, increasing energy security of the U.S. economy and military forces. DOD should play an active role in catalyzing the development of this infrastructure, and ensure that it will have the ability to use manufactured fuels in its future vehicles.

This page intentionally left blank

Mid-to-Far Term Fuel Strategy (1)

- *Liquid hydrocarbon fuels have ideal properties and are needed as transportation fuels for the foreseeable future*
  - *Oil-derived fuels primarily imported and will become increasingly scarce*
  - *Existing refinery infrastructure*
    - *Predominantly coastal and vulnerable*
    - *Operating at capacity*
- *Alternative: Fuel efficiency, domestic resources, interior production*

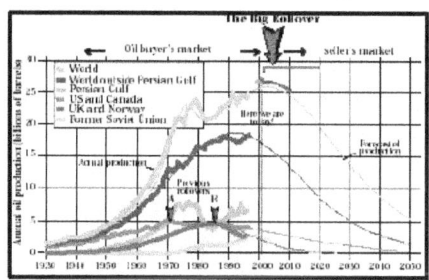

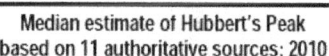

Median estimate of Hubbert's Peak
based on 11 authoritative sources: 2010

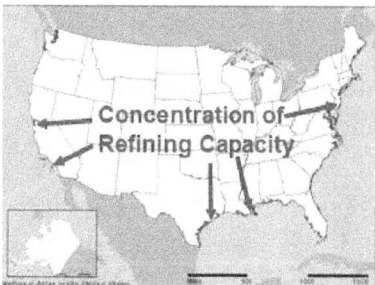

Concentration of Refining Capacity

Mid-to-Far-Term Fuel Strategy (1)

As stated earlier, the Panel's fact-finding sessions resulted in conclusions in three major areas – the third of which is the need for manufactured fuels.

As discussed earlier in the report, liquid hydrocarbons have ideal properties (especially volumetric energy density) as transportation fuels and will be needed by US military forces for the foreseeable future. Currently, such fuels are overwhelmingly obtained by refining crude oil. Already, the majority of US fuel is produced from imported oil, making the United States dependent on overseas production infrastructure (frequently located in politically unstable or unfriendly regions); a complicated transportation system; and a domestic crude oil refining infrastructure now operating near capacity. With regard to the last point, it is important to note that the US refining infrastructure is concentrated in a few coastal areas. Although there are inland refineries, they are typically associated with depleted oil fields; the coastal refineries are associated with crude oil imported from overseas, and delivered to the US via seagoing tankers. These areas are particularly susceptible to disruption from both natural disasters (earthquakes in California and hurricanes along the Gulf coast), and attacks specifically intended to disrupt US energy supply.

As developing economies in Asia rapidly increase their consumption of oil-derived hydrocarbon fuels, they will be competing with the United States, which now dominates world oil consumption. Such competition will drive prices ever higher, and perhaps lead to intermittent fuel shortages as production fluctuates. Clearly, this competition for resources also provides oil producers multiple options for selling their products, and raises the possibility that the United States could face shortages resulting from shifts in political alignments within the producing nations.

All of the points above are exacerbated by the inevitable exhaustion of crude oil, which is a finite resource. Geophysicist M. King Hubbert predicted in 1956 that US oil production would peak in 1970; this prediction was scoffed at, but proved to be remarkably accurate. His arguments took full cognizance of the fact that rising oil prices would motivate the extraction of additional oil from "exhausted" fields when the economic picture changed. Currently, there is a wide range of estimates for when "Hubbert's Peak" will occur for world oil production. In general, commercial oil companies tend to place the peak farther out in time, whereas Government and academic sources estimate the peak will be soon. Eleven authoritative predictions (subsuming all of these categories) yield a median estimate of 2010. Basing US economic and military use of liquid hydrocarbon fuels exclusively on crude oil feed stocks will therefore become problematic in the near future.

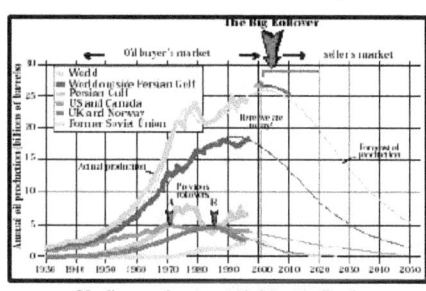

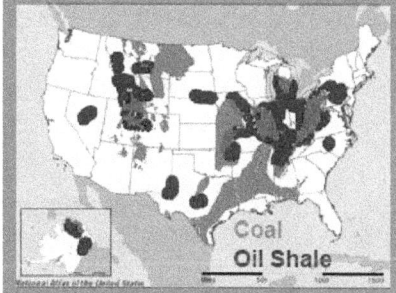

Median estimate of Hubbert's Peak
based on 11 authoritative sources: 2010

Mid-to-Far-Term Fuel Strategy (2)

Although increasing fuel efficiency will remain an important consideration in any scenario, there are limits to what can be achieved by this approach alone. An ideal future fuel strategy would couple conservation with exploitation of domestically abundant, non-crude-oil feedstocks. Fortunately, the US is blessed with huge reserves of coal and shale oil, and Canada has large reserves of tar sands. To date, all of these resources are underutilized. Much US coal is considered undesirable for electric power production because of high sulfur content, which leads to acid rain or requires expensive exhaust gas remediation; long-term contracts for this coal have remained very stable in price, unlike crude oil and natural gas. Shale oil and tar sands are under early stages of commercial development for liquid hydrocarbon fuel production, but are primarily used to supplement crude oil refinery products through blending, because conventional refining of these feedstocks produces long-chain hydrocarbons that are unsuitable, by themselves, as transportation fuels.

If acceptable transportation fuels could be produced from domestic feedstocks, the distribution of coal and shale oil could lead to geographically more dispersed refining infrastructure, which would decrease the susceptibility of fuel production to either natural disasters or deliberate attack.

This page intentionally left blank

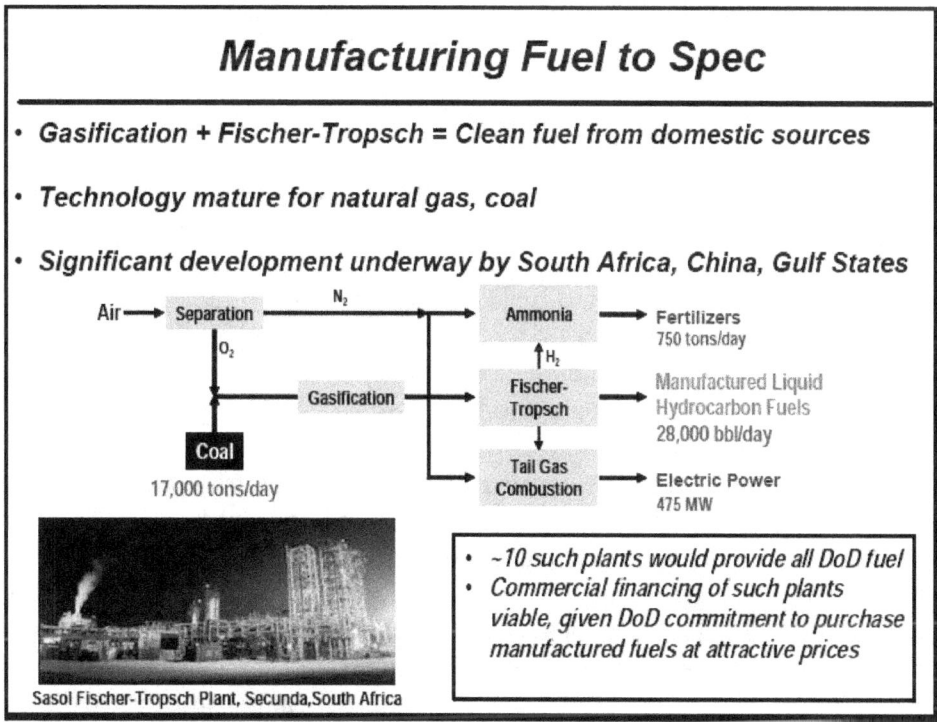

Manufacturing Fuel to Spec

The United States' future dependence on liquid hydrocarbon fuels without abundant domestic crude oil supplies will not be unprecedented. In pre-WWII Germany, Franz Fischer and Hans Tropsch developed a process to produce liquid hydrocarbon fuel from coal; the so-called Fischer-Tropsch (FT) process supplied a substantial fraction of Germany's transportation fuels, particularly after Allied actions threatened the output of the Ploesti oil fields and refineries.

In the FT process, so-called *syngas* (a mixture of molecular hydrogen and carbon monoxide) is reacted at high temperature in the presence of an iron catalyst to produce a mixture of short- and medium- and long-chain hydrocarbons, carbon dioxide, water, and hydrogen. The short-chain hydrocarbons (so-called *tail gas*) are not ideal transportation fuels, but can be burned locally to produce the necessary heat for the FT reactions, and can additionally be used to produce electricity from a gas-turbine generator. The medium-chain hydrocarbons are usable transportation fuels, particularly when blended with additional material derived from the long-chain hydrocarbons (usually waxes) through hydro-cracking. The ability to control the carbon chain lengths derived from waxes allows for the manufacture of ideal transportation fuels such as diesel and jet fuel.

Syngas is easily produced via the partial combustion of coal, which has been gasified and combined with molecular oxygen derived from air. Syngas (then known as water gas) was produced and distributed to homes and businesses in the late 1800s and early 1900s, before methane supplanted it for safety reasons (the carbon monoxide in syngas made it a very dangerous material). Today, gasification is usually accomplished with pulverized coal and pure oxygen produced by separating air (the nitrogen can either be vented to the atmosphere, or used with some of the hydrogen in syngas to produce ammonia, a nitrate fertilizer feedstock). Importantly, the gasification process serves to separate the sulfur and heavy-metal contaminants found in low-grade coal (which makes it undesirable as a raw fuel). Thus, the liquid hydrocarbon fuels produced from coal via gasification and the FT process are intrinsically clean. Use of such fuels will minimize emissions (sulfur and particulates) from internal combustion engines, and will also allow production of clean hydrogen (via fuel reformers) that could supply a fuel cell without poisoning the fuel cell chemistry.

FT fuel production is mature technology. As mentioned above, it was used successfully by WWII Germany on a large scale. South Africa was unable to import crude oil in large quantities during the apartheid era, and consequently all of South Africa's vehicles have been powered by FT-generated fuels derived from low-grade coal for *nearly fifty years.* Sasol's FT plant in Secunda, South Africa produces 150,000 barrels of manufactured fuel per day. China, which also has abundant domestic coal, has purchased essentially the entire world output of coal gasifiers for the past several years to produce fertilizer via the FT process. Finally, commercial oil companies are planning on establishing FT infrastructure in the Persian Gulf to produce liquid hydrocarbon fuel from natural gas (which would otherwise be flared off, or liquefied and transported to Liquefied Natural Gas (LNG) terminals at high expense in pressurized tankers). According to Shell Oil, by 2015 the Gulf State infrastructure will produce 900,000 barrels/day of FT-derived liquid hydrocarbon fuels from natural gas. This also points out that the FT process can be used to produce liquid hydrocarbon fuel from virtually any carbon-containing feed stock, including low-grade tars, biomass, or shale oil; only the preprocessing steps would differ from the gasification process used for coal.

The flow chart on the previous page shows an integrated gasification-FT-fertilizer-power plant proposed by Baard Generation (a 20-year-old producer of small- to medium-scale project-financed power plants). From 17,000 tons/day of low-grade coal, the plant would produce 28,000 barrels/day of liquid hydrocarbon fuel, 750 tons/day of ammonia, and 475 MW of net electrical power. The plant would cost $3B, and employ 200 full-time staff. Baard envisions building such plants near rich low-grade coal fields, areas that are typically economically depressed since emission controls have made such coal economically unattractive for power production. Although such plants are relatively small, it would only take about 10 such plants to supply all of DOD's liquid hydrocarbon fuel requirements. Baard claims that commercial financing of such plants will be possible, with adequate internal Return on Investment (ROI) and revenue/debt margins. DOD could, however, catalyze commercial development of this highly desirable infrastructure by making a long-term commitment to purchase liquid hydrocarbon fuels at attractive prices. Baard estimates that a 10-year commitment would enable a sale price (for diesel) of $61/barrel (bbl) ($1.45/gal); similarly a 15-year purchase commitment would yield $54/bbl ($1.29/gal) diesel. Over the term of the commitment, diesel prices would escalate, but only at the rate of long-term coal contracts, not at the rate of oil markets.

At such prices, given the impending arrival of Hubbert's Peak, DOD risks little by making a purchase commitment. In fact, long-term purchase contracts for FT-derived liquid hydrocarbon fuels could provide a highly favorable hedge against spot market prices for fuel.

In fact, the 2005 Energy Bill forwarded by Congress to the White House on August 3 (and signed by the President the week of August 8, as this report was being written), provides the authority for the DOD to procure alternative fuels produced from domestic supplies of coal, oil shale, and tar sands. In an amendment to Chapter 141 of title 10 of the United States Code, an new section inserted in section 2398 calls for the Secretary of Defense a) to develop a strategy to use fuels produced from the above-mentioned feedstocks to assist in meeting DOD fuel needs, and b) grants the Secretary of Defense multiyear contract authority to procure such fuels.

It should also be noted that Sec. 417 of the Energy Bill calls for the development of a very small FT plant specifically for the investigation of FT transportation fuels produced from Illinois Basin Coal. The plant authorized by the Bill would only produce 500 gallons of FT fuel per day, and as such would do little to contribute to the DOD's energy needs. However, the FT-fuel produced could be used to test compatibility with DOD vehicles.

The DOD should additionally use the multiyear procurement authority recently granted by Congress to catalyze the establishment, through commercial financing, of much larger FT plants of the type proposed by Baard Generation.

This page intentionally left blank

Conclusions: Manufactured Fuels

- *Liquid hydrocarbon fuel production using domestic energy sources is feasible*

- *Commercial financing and infrastructure development will drive this process*

- *DoD action needed to catalyze development & ensure US military takes advantage of manufactured fuels*

- *Need to ensure military platforms can use manufactured fuels*

**Manufacture Fuel from Domestic Sources —
Decrease Dependence on Imported Crude Oil**

Conclusions: Manufactured Fuels

In summary, the United States is in the fortunate position of having domestic resources that will, with the development of appropriate infrastructure, enable the continued use of liquid hydrocarbon fuels, without the economic and security disruptions attending to the import of crude oil as the primary feed stock. This change in posture need not be funded by the Government (and indeed, to realize the full potential of this approach, the Government could not afford to capitalize the needed changes in infrastructure); the rising price and increasing scarcity of crude oil will motivate commercial firms to invest in manufactured fuel infrastructure. However, the DOD can catalyze this process (while creating a valuable hedge for itself against future increases in oil prices) by committing to long-term purchase of manufactured fuels at attractive (even now) prices, under the terms of the recent Energy Bill

Further, since manufactured fuels lack many of the impurities of crude-oil-derived fuels, they will be more acceptable from the standpoint of emissions. However, some properties may differ from today's fuels. In particular, lubricity is strongly correlated with (otherwise undesirable) sulfur content; it will be important for the DOD to ensure that it performs the testing, and ensures that appropriate additives or other treatments are available to allow use of such fuels in the future tactical vehicle fleet. FT fuels produced from small experimental FT plants, like that proposed by the recent Energy Bill, will be useful for this purpose.

This page intentionally left blank

> # *Recommendations*
>
> ### *Nearer-term Payoff (PR 07/POM 08)*
> - *Fuel tether is still there, but...*
>   - *Found a way to lengthen it (HEVs)*
>   - *And untangle it (Fuel Management)*
>
> - *Commit to HEV technology for all future TWV*
>   - *Establish an HEV development roadmap*
>   - *Immediately initiate system engineering trade-offs*
>   - *Invest in on-going HEV development projects*
>
> - *Develop prototype system to enable real-time, in-stride fuel allocation for the Operational Commander*
>
> ### *Longer-term Payoff (2015 & beyond)*
> - *DoD catalyze manufactured liquid hydrocarbon fuels infrastructure*
>
> - *Characterize the compatibility of manufactured liquid hydrocarbon fuels with DoN equipment*

Recommendations

Nearer-Term Payoff (PR07/POM '08)

In response to Gen Mattis' challenge -- "Unleash us from the tether of fuel" -- the Panel determined that the tether remains but found a way to lengthen it (HEV technology) and untangle it (dynamic fuel management). Hybrid electric drive vehicles offer the most effective and efficient way to "unleash us from the tether of fuel." Improved fuel economy, as much as 20% or more, can significantly reduce shortfalls in fuel as well as reduce the expeditionary footprint. They enable highly maneuverable and agile vehicle traction control whether on or off-road, in covert or overt operations and can provide mobile electric power. They also offer additional trade-offs in reach and mobility as it relates to systems capability. To achieve this, however, a hybrid electric road map must be developed that will leverage commercial sector power electronics and control, energy storage and conversion, to include engine and fuel cells. Additionally, the military must continue developments in high torque motors, series architectures, control algorithms, energy storage and pulse power technology. With those developments, however, comes the need to initiate systems engineering trade-offs and investment in hybrid electric technology.

Presently the Marine Corps, as well as the Army, do not have the ability to effectively and efficiently manage fuel during combat operations. As operational reach is extended, accurate planning tools, real-time vehicle level fuel status, and location data indicators are critically needed to enable dynamic retasking of fuel assets on the battlefield, and thus, provide the ability to conserve fuel, sustain op tempo and reduce fuel train vulnerability.

Longer-Term Payoff (2015 & Beyond)

Although numerous alternative fuels are being evaluated across the spectrum of power and energy density to satisfy tomorrow's fuel needs for the United States, only liquid hydrocarbons can provide DOD with the properties needed for its transportation fuels in the foreseeable future.

Currently, these fuels are obtained from refining petroleum, but as has been discussed, these resources are waning and must be replaced with a suitable substitute. Fortunately, the United States has large deposits of coal and shale oil as well as large tar sand deposits in Canada. If developed, and it would appear that this is becoming more and more possible both economically and operationally through gasification and the FT process, it would greatly reduce this Nation's dependence on crude oil and enable DOD to have the fuels necessary for tomorrow's conflicts.

The Panel strongly believes that DOD needs to commit now to procuring manufactured liquid hydrocarbons for the long term at lower than current market price to encourage commercial financing, push technology and help motivate the building of the necessary manufacturing and distribution infrastructure.

Since manufactured liquid hydrocarbon fuels also lack many of the impurities of crude oil derived fuels, they will have better emission properties. With them, however, comes the need to develop the necessary additives and treatments to ensure engine compatibility. The DOD and DoN must, therefore, start now to develop the means to operate these fuels in legacy as well as future tactical vehicles, equipment and systems.

Actions (1)

- **Commandant of the Marine Corps (CMC)**
  - *Support application for emissions waiver submitted by Army*

- **ASN (RDA)**
  - *With Services, advocate the use of multiyear procurement authority granted SECDEF in 2005 Energy Bill to <u>catalyze commercial financing</u> of large-scale FT plants producing transportation fuels*

- **CG MCCDC (Request of CNR via CMC)**
  - *Establish new program elements (6.2 & 6.3) for HEV technologies*
  - *Demonstrate technologies for real-time fuel asset visibility*
  - *Develop real-time dynamic fuel allocation prototype system*
  - *Develop conditioning technologies for substandard tactical fuels*

Actions (1)

It is essential to the performance of current and future ground combat equipment that the waiver on emissions standards the US Army is seeking from the EPA is granted. The 2010 emission standards being imposed on the civilian sector would greatly reduce the performance of all systems with an internal combustion engine as well as negatively impacting the designs of future systems. Thus the committee recommends that the Commandant of the Marine Corps (CMC) take action to endorse the waiver the US Army is seeking on the grounds that the 2010 emission standards would adversely affect the ability of the Marine Corps to accomplish its national security mission.

Both the opportunity and need to utilize manufactured fuels is now present. The technology necessary is well proven and has been in operation in several countries for decades. The commercial (vice DOD) investment necessary to construct facilities is also available. As the percentage of imported fuels used by the United States steadily rises it has become a national security issue, both economic and military, that domestic sources of fuel be expanded.

The Panel recommends that ASN (RD&A) take the lead in developing a strategy among the Services to foster the commercial development of manufactured fuels along the lines discussed in this report.

In order to extend the operational reach of Marine ground forces the application of new technology to provide greater fuel economy and better fuel management is essential. In addition, "fuel as found" in expeditionary environments needs to be made more usable and effective when provided to U.S. forces. The Panel recommends that the CG MCCDC take action to request that the CMC direct the Chief of Naval Research (CNR) to accomplish the following tasks:

- Establish new program elements specifically directed at hybrid electric vehicle technology development for use by Marine Corps ground forces.

63

- Demonstrate in the near term technologies that can be used for real time fuel asset visibility.

- Develop a prototype "system" for dynamic fuel allocation.

- Develop conditioning technologies that can alter substandard tactical fuels for effective use by the GCE.

Actions (2)

- **CNR: Support these CMC tech investment requests**
  - **Complete RST-V Technology Program**
    - *Transition from DARPA to ONR for final maturation*
    - *Develop on-the-fly mission profile selection technology*
    - *Transition Mature Design to CG MARCORSYSCOM*
  - **Complete On-Board Vehicle Power Program**
    - *ONR Transition to CG MARCORSYSCOM*
  - **Conduct real-time fuels status tech demos**
  - **Develop Commander's real-time dynamic fuel allocation prototype system**
    - *Coordinate with DARPA to establish a joint program*
  - **Develop technologies for conditioning expeditionary substandard tactical fuels**
  - **Monitor status of FT Plant authorized by 2005 Energy Bill**
    - *Use fuel produced to conduct research on compatibility with current and future TWVs*

Actions (2)

The ONR needs mature HEV technologies transitioned to Marine Corps programs of record. The CNR should initiate the following actions in response to the CMC:

- Complete the RST-V technology demonstration program after the currently funded DARPA project ends.

- Develop the on-the-fly mission profile selection technologies for use by tactical HEV systems.

- Complete the ongoing ONR funded On-Board Vehicle Power Program

- Transition all technology to CG Marine Corps Systems Command (MARCORSYSCOM) as soon as practical for implementation in the POM-08 time frame

The CNR will also respond to the CMC on accomplishment of the following tasks:

- Conduct real-time fuels status technology demonstrations responding to Marine Corps established operational criteria

- Develop a prototype, potentially as a joint program with DARPA, for the Commanders Real Time Dynamic Fuel Allocation System

- Develop technology for conditioning "found" substandard expeditionary fuels. Monitor the status of the FT Plant authorized by the 2005 Energy Bill, and use fuel produced by it to conduct research on compatibility with current and future TWVs

This page intentionally left blank

Appendix A
Terms of Reference
Future Fuels

Objective

Identify, review, and assess technologies for reducing fuel consumption and for militarily useful alternative fuels, with a focus on tactical ground mobility. Technical maturity, current forecasts of "market" introduction, possible operational impact and S&T investment strategy should be considered. Two main focus areas to be considered in this effort are alternative fuels, and improving fuel efficiency (to include examination of alternative engine technologies).

Background

Lessons learned from OEF/OIF, as well as analyses of future warfighting concepts such as Enhanced Networked Seabasing and Ship-To-Objective Maneuver, have identified fuel consumption and distribution as being among the most critical aspects of projecting and sustaining a combat credible force. Future naval forces must have a secure fuel supply to be effective. Innovative approaches must be devised for the responsive and flexible delivery of required fuels to ground, surface and air forces maneuvering throughout the battlespace.

This effort is designed to focus on the consumption part of that equation. Alternative fuels offer potential not only for ground, sea and air vehicles but also for fuel cells to power portable electronic systems. Marine Corps operating forces need long-shelf-life, high-capacity, longer-lasting, lightweight, renewable, environmentally-friendly, multi-application energy sources.

The technical maturity of future alternative fuels and conversion technologies must be objectively quantified to better understand not only the realistic capabilities of each technology and the probable timeframe in which it can practically be deployed, but to guide the Navy's S&T investment strategy in relationship to other industry and government organizations. Establishing an effective strategy that could optimize/leverage the cooperative research among industry, DoE, DOD and other government organizations should be considered.

Specific Taskings

This study will specifically:

- Identify, review and assess technologies for fuel consumption reduction, to include alternative engine technologies,

- Identify alternative fuels and assess readiness for introduction to Naval forces,

- Evaluate relevant Commercial-Off-The-Shelf (COTS) technologies for operational utility and suitability,

- Identify candidate high-payoff S&T areas for further study, development and fielding by naval forces,

- Recommend guidelines for establishing an effective strategy that could optimize/leverage the cooperative research among DOD, DoE, and industry.

This page intentionally left blank

Appendix B
List of Briefings

Ms.Jan Gnerlich	ONR
Dr. Rick Coffin	NRL
Dr. McGrath	ASN
CWO Kunneman	HQ, USMC
Dr Alan Roberts	N-42
Mr Bill Harrison	OSD
Mr. Anthony Nickens	ONR
Mr. Joe Pizzino	NSWC CD
LtGen James Mattis	MCCDC
Mr. Al Sawyers	MCCDC, S&A
Mr. Cortez Stephens	MCCDC, S&A
Mr. Dennis Hardy	NRL
Dr Bhakta Rath	NRL
Dr. Rich Carlin	ONR
Dr. Jerry Hu	ORNL
Dr. Frank Rose	Board on Army S&T (NRC)
Col Rohrer	DLA/DESC
Dr Freeman	HQDA (S&T), ASAALT
Mr. Dan Herrera	TARDEC
Mr. Charles Raffa	TARDEC
Mr. Ed Shaffer	ARL
Dr. Peter Schihl	TACOM
Dr. David Beach	ORNL
Dr. Szostak	DARPA
Mr. Philipp Patch	MCCDC, EFDC
Mr. Angus Hendrick	NAVSEA
Mr. Rick Kamin	NAVAIR
Mr. Steve Nimmer	Oshkosh Trucks
Mr. Terry Goff	Caterpillar
Mr. John Hartranft	NAVSEA

Mr. John Baardson	Baard Generation
Mr. Charles Lucius	Battelle
Dr. Walter Bryzik	TARDEC
Dr. Ron Graves	NTRC
Mr. Terry O'Connor	Shell Oil
Mr. Michael Collins	NAVSEA
Mr. Joe Tomita	Toyota
Mr. Michael Cunningham	TRADOC
Dr. Tim Armstrong	ORNL
Mr. Mark Ouellett	Army, PM Unit of Action
Dr. John Pazik	ONR
Prof. Rolf Reitz	Engine Research Center, Univ. of Wisconsin
Dr. JoAnn Milliken	DOE
United Kingdom:	
Dr. Lee Juby	Rolls Royce
Mr. Rob Hughes	Rolls Royce
Mr. Carl Bourne	Rolls Royce
Mr. Paul Maillerdet	UK MOD
Mr. Nigel Johnson	DSTL Land Systems
Mr. Stuart Burdett	DSTL Land Systems
CDR John Wood	UK MOD DPA

Appendix C
Panel Membership

Panel Membership

The study sponsor is Lt.Gen James N. Mattis of the Marine Corps. He commanded the marines on their dash to Baghdad. That experience and prior responsibilities spawned this important study. The executive secretary is a Marine Corps Major, a helicopter pilot with logistics experience. An outstanding panel of experts was assembled to address this complex and challenging problem. The Panel members brought years of valuable experience in both operations and acquisition with retired senior officers from the Marine Corps, the Navy and the Coast Guard. These senior military officers together with leading technology authorities from the ranks of former government technology officials, industry, national labs, service labs, OSD, academia and university-affiliated research centers brought their rich experience to develop the recommendations and actions for this study.

This page intentionally left blank

Appendix D
Acronyms

AAV	Amphibious Assault Vehicle
AHED	Advanced Hybrid Electric Drive
AOR	Area of Responsibility
ASN	Assistant Secretary of the Navy
CFR	Code of Federal Regulations
CG	Commanding General
CMC	Commandant of the Marine Corps
CNR	Chief of Naval Research
DARPA	Defense Advanced Research Projects Agency
DASN	Deputy Assistant Secretary of the Navy
DOD	Department of Defense
DOE	Department of Energy
DON	Department of the Navy
EGR	Exhaust Gas Recirculation
EPA	Environmental Protection Agency
FT	Fischer-Tropsch
GCE	Ground Combat Element
GDLS	General Dynamics Land Systems
HEMTT	Heavy Expanded Mobility Tactical Truck
HEV	Hybrid Electric Vehicle
HMMWV	High Mobility Multi Purpose Wheeled Vehicle
HQMC	Marine Corps Head Quarters
IPS	Integrated Propulsion System
LAV	Light Armored Vehicle
LNG	Liquefied Natural Gas
LVS	Logistics Vehicle System
MARCOSYSCOM	Marine Corps Systems Command
MCCDC	Marine Corps Combat Development Command
MEF	Marine Expeditionary Force
MEP	Mobile Electric Power
NRAC	Naval Research Advisory Committee
NRL	Naval Research Laboratory
OIF	Operation Iraqi Freedom
ONR	Office of Naval Research
OPNAV	Office of the Chief of Naval Operations
OSD	Office of the Secretary of Defense
POM 08	Program Objective Memoranda 2008
PR 07	Program Review 2007
RD&A	Research, Development and Acquisition
ROI	Return on Investment
RST-V	Reconnaissance Surveillance Targeting Vehicle
S&T	Science and Technology
TOR	Terms of Reference
TWV	Tactical Wheeled Vehicle

This page intentionally left blank